Science Fact or Science Fiction?

Answering children's science questions at Key Stage 1

John Stringer

QUESTIONS
PUBLISHING

First published in 2000
by The Questions Publishing Company Ltd
27 Frederick Street, Birmingham B1 3HH

Edited by Diane Parkin
Designed by Al Stewart
Illustrations by Sue Woollatt, Graham-Cameron Illustration
Cover design by Lisa Martin

ISBN: 1-84190-024-9

Activities in this book have been selected with care, but safety advice must be followed. Neither the author nor the publisher shall be liable for injuries that may be caused by not following the instructions.

Also available from The Questions Publishing Company Ltd:
Science Fact or Science Fiction?
Answering children's science questions at key stage 2

Contents

Introduction: science fact and science fiction

Before you read any further, answer these questions from your own knowledge and understanding:

1 Trees are big things. When we cut them down we find that they are made up of a lot of wood. But where did all that wood come from? What were the raw materials? How was it made?

2 If you cool a liquid, it will become solid and that solid will be smaller than the liquid it came from. So why does frozen milk push the top off the bottle?

3 Is the dark side of the Moon really dark?

We all have our own ideas of the way the world works. Children are no exception. Much recent research has been devoted to understanding children's ideas about the way the world works. This constructivist research has led us to a clearer understanding of the way children think. And it has led to a style of teaching science to young children that begins by challenging their prior ideas.

This approach can be effective in changing children's thinking so it is closer to the accepted scientific explanation. You might say, 'so that it is closer to the truth', but that is to assume that we know the truth – and it would be fair to say that scientific understanding is always developing with new explanations of what we originally believed we all understood. Nevertheless, there are explanations that are nearer right than others and these explanations stand up to scientific scrutiny and rigorous testing.

Teaching to what children already know is both sensible and practical. But it can be undermined by the simple fact that many teachers have their own prior ideas about the way the world works – and sometimes these ideas do not match the accepted scientific explanation either. There is no shame in this. Lewis Wolpert, a professor of science, has argued that science is not about common sense. If you want to understand science, he says, leave your common sense at the door. Most primary teachers survive by common sense. Tough. It is no use to you in science. In fact, if you give a child a common sense answer, you are probably wrong.

Take our three starter questions. What were your answers?

1 Trees are big things. When we cut them down we find that they are made

up of a lot of wood. But where did all that wood come from? What were the raw materials? How was it made?

If your answer was that the tree somehow sucked all that material from the ground, then you have to explain why trees aren't standing in great big holes. No, the tree, all its leaves, its trunk, branches, flowers and fruit, were all once water and carbon dioxide. By a miraculous process we call photosynthesis the tree (and other green plants) harnessed the energy of the Sun to convert these raw materials to both its structure and the energy it needs to live. (And incidentally, poured oxygen into the atmosphere that you and I need to breathe and live.) Sure, it uses tiny amounts of mineral salts, and we may give it extra, misleadingly calling this plant food' The lesson? Expect the unexpected.

2 If you cool a liquid, it will become solid and that solid will be smaller than the liquid it came from. So why does frozen milk push the top off the bottle?

Right. If molten metals cool they become solids that take up less space than they did as a solid. The same is true of other materials that exist in both solid and liquid forms. But water – the most abundant liquid on the planet – breaks the rules. Ice cubes are larger than the water they were made from. Frozen coolant can break a car's engine block. And milk – which is mostly water – will expand out of a bottle and force off the cap.

Good thing too. Because it is bigger and less dense than water, ice floats. If it didn't ponds would freeze up from the bottom and there would be no fish in them. Even more importantly, there would never have been life on Earth as we know it. Life originally evolved in the water. And even if it had begun to develop, it would have had to survive periodic freezing, or maybe being brained by sinking ice cubes.

The lesson? The best rules can be broken.

3 Is the dark side of the Moon really dark?

No it isn't. There are times each month when the Moon is only slightly lit on the face – the side that constantly faces us. On these occasions the other side of the Moon is in sunlight. The dark side is light. The word is inaccurate.

The lesson? Don't trust the words we use.

Still with me? Then it's time to try applying all this. Turn to *How to use this book*.

How to use this book

The book is divided into units, each one referenced to the National Curriculum for Science 2000. There are 17 units at Key Stage 1. There are 19 in the Key Stage 2 book.

Each unit is titled by a common question or misconception. Then sections follow that enable you to challenge children's thinking and work towards the learning outcome.

This is the structure of each page:

Key Stage 1

National Curriculum reference: This is to the new National Curriculum for Science 2000.

What children may think

A few lines cannot do justice to the research of the constructivist projects, which include CLiS (Children's Learning in Science) and SPACE (Science Processes and Concepts Exploration). These should be read for themselves.

This paragraph aims to give you a sense of the kind of prior ideas that children may have. Challenging those ideas may demonstrate to you how tenaciously they are held.

What you may think

Most of us have our own prior ideas on science, made up of incompletely understood school experiences, popular science and rationalisations. For examples of this, see the Oxford University PSTS (Primary School Teachers and Science) project. This paragraph may challenge your own thinking.

The accepted science

This section presents the commonly held views, defines a few terms, and reminds you of some of the conventions of science.

The photocopy sheets

Each unit has two photocopiable resource sheets which you may choose to use – in whole or in part – or to modify to your children's needs.

Some are practical, some pencil and paper. Where appropriate, answers are given.

Some more activities

This section lists other activities that you could call into play in your teaching.

Learning outcomes

This section lists what most children should know and understand following your teaching, in 'Childspeak' form.

Part 1

Sc1 Scientific enquiry

Key Stage 1
National Curriculum reference: Sc1, 1a, 2a, 2b

1 How important is evidence?

What children may think

I'm doing this investigation. I'm growing this plant in the cupboard. You'd think it wouldn't grow there in the dark. But it has! It's grown all yellow and straggly. So my investigation has gone wrong. I'm not going to say that. I'll write "my plant did not grow". That'll do it.

What you may think

Evidence is important. Children need to collect it from their investigations. But a lot of investigations do go wrong. That's really frustrating. When I learned science, at least we knew what we were trying to find out. It was there at the top of the board. There were no surprises, either. We got the results the teacher wanted – or we were wrong.

The accepted science

Science is one way of finding out more about the world. The strength of science is the scientific method, the collection of evidence from investigations. Investigations don't always give you the answer you want – or expect. This doesn't mean the enquiry has gone wrong.

For example, you might decide to investigate the germination of seeds. You might reason that since plants need light they are unlikely to germinate in the dark. But you find that the seeds not only germinate, they produce plants that are longer (though less healthy) than plants on the window sill. But the enquiry has not gone wrong. Most seeds germinate underground in the dark. Plants need light and if you don't find light underground, then the best thing is to grow towards it. Hence the results confirm what happens in the world.

The photocopy sheets

1 The completed table should look like this:

Science questions	*Will you look around you?*	*Will you look in a book?*	*Will you do an investigation?*
1 What colour are your friend's eyes?	Yes		
2 Who was the first man on the Moon?		Yes	
3 What is the classroom floor made from?	Yes		
4 Are all metals magnetic?		Yes	Yes
5 Why do plants need water?		Yes	(You could investigate, and you would find that plants die without water, which does not answer the question).

2 Provide two magnets, one large and one small. You will frequently find that small Alnico magnets are stronger than large traditional magnets. Children usually expect the opposite.

Children will have many ideas. Watch out for the fallacy that a strong magnet will lift a weak one. Both are pulling. Don't end with the misconception that small magnets are always stronger that big ones.

Some more activities

Use every opportunity to examine evidence and decide how valid it is. Can we believe the evidence of our own eyes?

Learning outcomes

Know that science answers questions by investigation, and that sometimes those answers will be unexpected.

How important is evidence?

What you need
A pencil and this sheet

What to do
How will you answer these questions?
Will you look around you?
Will you look in a book?
Will you do an investigation?

Tick the boxes.

Science questions	*Will you look around you?*	*Will you look in a book?*	*Will you do an investigation?*
1 What colour are your friend's eyes?			
2 Who was the first man on the moon?			
3 What is the classroom floor made from?			
4 Are all metals magnetic?			
5 Why do plants need water?			

What to record
your answers in the table.

How important is evidence?

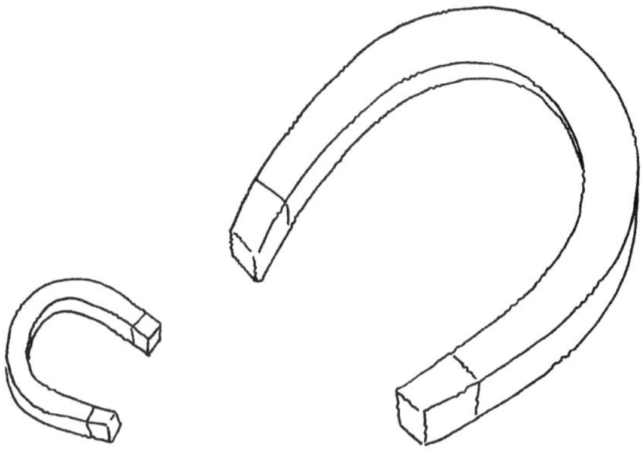

What you need
Two magnets - one big one small

What to do
Investigate.
Find out which is the stronger magnet.
Guess first.
Then try it.

What to record
What did you do?

What did you find out?

Key Stage 1
National Curriculum reference: Sc1, 2a, 2c

2 How do you predict the outcome?

What children may think

We've been asked to guess what will happen when we put this jelly cube in water. I don't know because I've never made a jelly. You get jelly in moulds don't you? I suppose if you put this cube in water it might disappear. It'll disappear if I put it in my mouth. It might colour the water too. I don't really know.

What you may think

It's hard to get children to ask questions these days. So much is given to them. They see it all on television. Even if they ask questions, we can't answer them in the classroom.

Prediction is important. Children should have a guess at what they think is going to happen. But little children have so little experience and they often expect something really exciting to happen. What they get can be a bit dull.

The accepted science

Science is about questions not about answers. Even when you've done an enquiry, and even if you get the accepted answer, you will still be left with questions.

Most questions cannot be answered by classroom investigations. Many are matters of fact and you can find out the answers from books or other sources. Some could be investigated if you had super science facilities. Only a few can be answered by classroom enquiry.

You need some sort of experience to begin to predict. Until you've watched what happens to a jelly cube put in water, how can you guess how to dissolve it quickly? The simplest experiences – sometimes play experiences – can fascinate the most sophisticated children.

The photocopy sheets

1 Provide ice cubes and saucers or other containers to hold the melt water.

Children enjoy watching the cube melt. Compare their predictions (guesses) with what really happens. The whole thing can be speeded up by putting the ice cubes on a safe water-filled radiator.

2 Children can suggest ways of dissolving jelly cubes quickly after they have had experience of dissolving a cube in water. Look for ideas about stirring, warming, even cutting the cube into small pieces.

Some more activities

Always ask for predictions before any science activity. Compare predictions with outcomes.

Learning outcomes

Know that a prediction can be very helpful in planning an investigation.

How do you predict the outcome?

What you need
An ice cube.
A saucer.

What to do
Watch the ice cube melt.
Guess which bits will melt first.:
the top?
the bottom?
the edges?
the corners?

What to record
Draw two pictures.
Draw one of your guess.
Draw one of what happens.

My guess

What I saw

How do you predict the outcome?

What you need
A jelly cube.
Some water.

What to do
Guess what happens when you put the jelly in the water.
Try it.
Watch.

What to record
You have to dissolve a jelly cube quickly.
What will you do?
Draw or write your ideas.

Dissolving a jelly cube quickly.
Draw or write your ideas:

Key Stage 1
National Curriculum
reference: Sc1, 4d

3 How could you get better?

What children may think

We've only just finished our experiment and now she asks us how we could make it better. Well we can't make it better. It worked didn't it? The little parachute came down really quickly. The big one took ages. Mark's soldier was bigger than mine, of course, and he didn't climb as high as I did. And his handkerchief was dirty. Yuk!

What you may think

It's tough enough getting children to do investigations without asking them how good they were. Besides, I know the answers. Karen's group is always a bit uncertain in science and they're bound to say that they weren't any good, and it didn't work, and they could do it better another time. But I'm not going to make them am I? Bart's group will say the opposite, that their experiment was absolutely perfect and there was no way they could improve on it and anyway they're bored and can we move on to something new?

The accepted science

It is important to review any investigation. Only by reviewing can you improve. Only by reviewing can you test the strength of the evidence. Is it good enough to say that you can expect the same results every time? Can you draw any general rules from it?

But objective judgements of your own performance are always difficult. It can help to give yourself a score. How did that investigation go? Give it a mark out of ten. Can you justify that nine, Bart, when your test wasn't fair and you made up the answers? Why only five, Karen, when you got the same result three times running?

The photocopy sheets

1 This is about samples and jumping to conclusions. Until Peter looks further in the garden he cannot tell whether snails like walls or wet weather. Children should be suggesting that Peter will find snails in other places, he just needs to look. His investigation is not wide enough.

2 Debbie knocks her rain investigation over and fiddles the results. This often happens when things go wrong. But it doesn't give right answers.

Some more activities

Ask the children to look at the ways in which the others did investigations. How could they learn from the work of others?

Learning outcomes

Know that reviewing an enquiry helps judge the value of the results and leads to better skills next time.

How could you get better at drawing conclusions?

What you need
This sheet and a pencil.

What to do
One wet day Peter found ten snails on the garden wall.
"Snails like walls", said Peter

What to record
Was Peter right?
Do snails like walls?
Where else should he look?
Why?

Write your answers here:

How could you get better at investigating?

What you need

'How much rain is falling? Debbie wondered.
She put a bucket out in the rain.
She caught the rain in the bucket.
She knocked the bucket over.
The rain water ran out.

'Never mind,' said Debbie. 'I'll fill the bucket from the tap. Then I'll measure it.'

What to do

Was Debbie right?
Does Debbie know how much it rained?
What should Debbie do to find out?

What to record

Draw what Debbie should do.

Part 2

Sc2 Life and living processes.

Key Stage 1
National Curriculum
reference: Sc2, 1a

4 How do you know it's alive?

What children may think

For many children movement signifies life. For this reason they may find it hard to recognise plants as living things. It is possible, but less likely, that they see other moving things – clouds, fire, even rivers – as living in some way. Given that children often personalise things, giving them personalities and becoming distraught at the loss of something inert, it is easier to understand why the living/never lived distinction is so hard to make.

What you may think

You may think you are pretty sure on what's living and what is not. But there are a number of living things in the everyday world – lichens and moulds, for example – that offer so few clues that they are living that you may be tempted to overlook them

It is also clear that the step from life to death is by no means always abrupt. People and other animals may go into a decline where it is no longer possible to identify the moment of death.

Finally there are some forms which may or may not be classed as alive. Viruses have no capacity for independent life, living instead inside the cells of a host. Outside the cell they may even crystallise.

The accepted science

There is a range of phenomena – growing, respiring, responding to stimuli, reproducing and, in the case of animals, moving and feeding – that identify a living thing. All living things can be shown to exhibit some, if not all, of these characteristics.

The photocopy sheets

1 Mouse, bird, tree, snail, child. Others may move but they are not alive.

2 Karen's plant will grow, change and flower. It may have fruits and seeds. It's alive. The children's drawings should show some of these activities.

Some more activities

Look around for examples of living things. Introduce one or two unusual ones – including safe fungi and yeasts.

Learning outcomes

Know that living things have characteristics that may include growing, respiring, responding to stimuli, reproducing and, in the case of animals, moving and feeding.

How do you know it's alive?

What you need
This sheet and a pencil.

What to do
Look at the pictures on this sheet.
Draw a ring round those things that are alive?

What to record
Choose one living thing.
How do you know it is alive?
Write your answer.

I chose:

I know it is alive because:

How do you know it's alive?

What you need
This sheet and a pencil.

What to do
Karen grew a plant.
She watered it every day.
'It doesn't move, she said. 'I think it's d

What to record
Show Karen that her plant is alive.
Write or draw some things that plants c
in this box.

Plants are alive because:

Key Stage 1
National Curriculum
reference: Sc2, 5a

5 How well do you know your environment?

What children may think

Children frequently know their local environment better than adults to some extent. After all, when was the last time you crawled on your hands and knees through the grass or scrambled up a tree? But they may have other less well-founded ideas. Most young children have difficulty in thinking about the environment as containing many interacting living things. They think rather of individual plants or animals. Many have difficulty in separating the environment from the people who live there. And they may consider that all the living things in the environment are fed by people.

What you may think

You may seriously doubt that there is 'environment' in or around your school. In many urban situations it is hard to think in terms of living things. Yet no environment is without life and it may be that the children can lead you to the ants' nest or hibernating hedgehog that can confirm this for you.

The accepted science

There is a difference between our uses of the words 'environment' and 'habitat'. The environment is fairly general. Our surroundings with its living things and made objects. A habitat is far more specific – an animal or plant address. Thus a habitat can be a pond or wood, a field or playground. Within a habitat plants and animals interact – eating or living off each other. This kind of relationship may also be called an ecosystem.

The photocopy sheets

1 The children are asked to find one plant and one animal in the environment. You can choose to let them bring the animal inside and to pick a common weed. They should say where they found them.

2 The children are asked to link the animals and plants to the correct habitat. You could discuss how they made their choices.

Some more activities

Compare the familiar environment of your school or the children's homes with somewhere more unusual.

Learning outcomes

Know that the plants and animals in the local environment depend on each other.

How well do you know your environment?

What you need
This sheet.
You might take a magnifying glass.

What to do
Go outside.
Find one plant.
Find one animal.
Draw them both.

What to record
Next to each picture, write and draw about where you found them.

My plant

My animal

Which is the right environment?

What you need
This sheet and a pencil.

What to do
Look at these plants and animals.
Some belong in a hot place.
Some belong in a cold place.

What to record
Draw lines to join the plants and animals to the cold place or the hot place.
Put them where they belong.
Explain why.

Key Stage 1
National Curriculum reference: Sc2, 2b, 2c

6 Who's on a diet?

What children may think

Most young children imagine that their body is a kind of skin bag that holds all food, blood and waste. Stomachs are commonly imagined to be large, low-slung containers, and they may have two exits, corresponding with the double functions of lavatories. But children will also have connected the stomach with the breaking down of food, and they may sense that it is transferred somewhere else in the body.

The word diet however is almost always associated with weight loss – and not with the balanced intake of foods that should characterise healthy eating.

What you may think

This depends on what you read last. Red wine is good for you/is bad for you; milk is a complete food/is fattening; all fats are harmful/some are essential to body functions. Advice changes day by day making teaching about balanced diet very difficult at any time.

The accepted science

We need a range of foods for healthy living. These can be provided from a variety of sources, and commonly include carbohydrates for energy, fats for both energy and general body functions, proteins for growth and repair, and small amounts of vitamins and minerals for general health and resistance to disease. We can obtain these foods from many different sources, including plant and animal. A balanced diet provides them in the right quantities.

The photocopy sheets

1 Doughnut Dan and Watery Dave should be shown as fat and thin. Talk about why, and challenge any ideas that diets are purely weight-reducing.

2 Children should ring – and explain – the best rabbit foods. Rabbit pellets contain balanced food. Note that lettuce can be harmful for baby rabbits.

Some more activities

Ask the children to record their food intake for a day. Discuss the outcomes remembering that many children have no choice in their diet.

Learning outcomes

Know that the right types and amounts of food are essential to stay healthy.

Who's on a diet?

What you need
This sheet and a pencil.

What to do
Dan ate only doughnuts.
Dave just drank water.

What to record
Draw Dan on his doughnut diet.
Draw Dave on his water diet.

Say why both were wrong.

Dave

Dan

The best diet for a rabbit

What you need
This sheet and a pencil.

What to do
Sharanjit has a rabbit.
Choose the food she should feed it.

What to record
Draw rings round the best foods for Sharanjit's rabbit.
Explain your choices.

Key Stage 1
National Curriculum
reference: Sc2, 3a

7 Why do plants need water?

What children may think

Why do we water plants? Children are unlikely to have thought about how plants grow. They may not even be aware of the daily and seasonal changes to plants or even that they are living things. Yet plants have this curious demand for water and children answer it in abundance – often drowning plants under gallons of water – when they don't neglect them altogether. It's a good tip to put a film canister next to a classroom plant as a 'watering can'. It helps prevent over-watering.

What you may think

There's no doubt that plants need water. They wither and may die if not watered. Plants draw this water from the ground and it travels through the stems to the leaves. You might conclude that water has some transport function, that it carries food around the plant and possibly waste. To some extent this is true. But it would be a wasteful transport system that drew gallons of water from the ground and transpired it by the bucketful without the water having any other use.

The accepted science

Water is a raw material of photosynthesis. Green plants can combine water and carbon dioxide to make sugars – the building blocks of growth and activity. Oxygen is a waste product of this process. The plant harnesses the energy of the Sun and involves a complex chemical called chlorophyll, which accounts for both the spread of leaves – to capture as much sunlight as possible – and their green colour.

So water is both a raw material and a means of transport – carrying the products of photosynthesis around the plant. (Remember that plants respire too – and need oxygen to release the stored energy of their home-made sugars.)

The photocopy sheets

1 The children are asked to choose from and explain their choice from:
Mark never watered his plant.
Helen gave her plant a little water every day.
Jane gave her plant lots of water.

They should recognise that plants need water but that over or under watering may be harmful.

2 The children are asked to:

Water your plant when it needs water.
Touch the soil.
When the soil feels dry, give the plant one canister of water.

They can record the water in canisters a day and record how the plant grows. Film canisters are commonly available from pharmacies.

Some more activities

Try growing plants with different needs for water – cacti, succulents and plants demanding regular watering.

What the children should know

Know that plants need water. It is essential to life.

Why do plants need water?

What you need
This sheet and a pencil.

What to do
Mark never watered his plant.
Helen gave her plant a little water every day.
Jane gave her plant lots of water.
Whose plant grew the best?
Why?

What to record
Circle the plant you think will grow the best.

Tick the true one:
Plants don't need water.
Plants need lots of water.
Plants need water.
Explain why.

Why do plants need water?

What you need
A plant in a pot.
A film canister.

What to do
Water your plant when it needs it.
Touch the soil.
When the soil feels dry give the plant one canister of water.

What to record
Record the water you give to the plant.
Record how the plant grows.
Make a table of how the plant grows.

Key Stage 1
National Curriculum reference: Sc2, 4b

8 What do you call this animal?

What children may think

The classification of plants and animals has developed over many centuries of study, and it is little wonder that children do not group living things in such a sophisticated way. They may, for example, take the first thing they are asked to classify as a criterion, and then compare all other examples with it. 'That's a beaver. It's an animal. That isn't an animal – that's a bird and that's a snake – it hasn't got fur like the beaver.'

They may not recognise that all the stages of a life cycle – a seed, tadpole or caterpillar – are in fact examples of one plant or animal. And they may have difficulty with sub-sets of groups – trees are flowering plants, but may not meet the common idea of flowers.

What you may think

Are you so good at this game? There are some pitfalls. The word 'animal' may be used interchangeably (and wrongly) with 'mammal'. We may use 'insect' for all creepy-crawlies, regardless of the number of legs. We can be confused by the everyday names for things – so jellyfish and starfish – both very different from fish and from each other – both become classified as 'fish'.

Are fungi plants? They have recently been put in their own group, separate from the plants. What about mosses and ferns?

The accepted science

It is worth looking at a general classification of plants and animals. It will remind you of the main groups. Animals are the large group that includes all invertebrates and backboned animals. Mammals have fur and give birth to their young alive. Insects are a very successful group of animals without backbones. They always have six legs (don't be confused by caterpillars that have additional false legs). Green plants are part of a special group that includes the trees and grass. Mosses and ferns do not have flowers and so are in groups of their own. And don't forget that a species can show variation within it. All may be dogs, but rottweillers, spaniels and a Chihuahua are very different.

The photocopy sheets

1 The children are asked to group the animals. They should end with three groups. Dogs and fish are easy but insects have six legs and a caterpillar has extra false legs.

2 The odd one out is:

Group of insects – fly, beetle, wasp and spider. Spiders have eight legs – not six. Group of flowering plants – dandelion, daffodil, rose – and toadstool. Toadstools are not plants. They can't make their own food.

Some more activities

Any naming activity. Use pictures, real living things or even animal sounds.

Learning outcomes

Know how to group living things by the way they look.

Where does this animal belong?

What you need
This sheet and a pencil.

What to do
Look at these pictures.
These are all animals.

What to record
Join the animals that belong in the same group with lines.
How many different groups have you got?

Explain your groups.

Which is the odd one out?

What you need
This sheet and a pencil.

What to do
Look at these pictures.
Some are plants.
Some are animals.
Each has an odd one out.

What to record
Circle the odd one out.
Explain why it doesn't belong.

Key Stage 1
National Curriculum reference: Sc2, 5b

9 Where do these plants and animals belong?

What children may think

Once upon a time there was a little green frog. He hopped along the road looking for a place to live. That wood won't do, he thought. Much too dry. That field is too sunny and that river is flowing much too fast. But that pond is just right. Perfect, he thought. And he hopped into the pond and lived there happily ever after.

This is a reasonable way for children to think about the way that plants and animals are adapted to their environments. And for some plants and animals there might be imagined to be an element of choice (compare the parable of the sower). Seeds that make the wrong choice won't grow.

What you may think

We mostly take for granted that we shall see particular plants and animals in particular circumstances. When we don't, it is a cause for surprise and curiosity: a rare bird off our coast; a flower growing in unexpected conditions; a panther in the Surrey countryside. This is the way things are. And if we have considered this at all it is in terms of common sense. Polar bears would have a hard time of it in the Sahara, camels in the Arctic.

The accepted science

Plants and animals have evolved to suit the place where they live. They are suited, by body structure or lifestyle, to living in those conditions. They can withstand the extremes that this habitat can throw at them – surviving drying out, or severe cold, high winds or long periods of darkness. Those that couldn't cut the mustard in these circumstances died out. Those that could passed their advantages to their offspring. There can be changes of unexpected severity. Ponds may dry up or forests be burned down. In these circumstances mobile animals can move elsewhere – possibly finding similar conditions. Maybe in the long run adapting to new habitats. Plants and immobile animals have to stay and sit the change out – a strategy which may be disastrous. Humans have made some enormous environmental changes. These changes have led to developments which have included adaptation for some plants and animals and extinction for many.

The photocopy sheets

1 The children are asked to put the animals back in the right places, and explain why.

2 The children are asked to put the plants back in the right places, and explain why.

Some more activities

Any outdoor experiences that show that plants or animals are found in a given place. 'Why are you looking there?' 'Why are they found there?' 'Why are you putting them back there?'

Learning outcomes

Know that animals and plants are adapted to suit the place they live.

Where do these plants and animals belong?

What you need
This sheet and a pencil.

What to do
Look at these pictures.
What is wrong with them?

What to record
Put each animal back where it belongs.

Where do these plants and animals belong?

What you need
This sheet and a pencil.

What to do
Look at these pictures.
What is wrong with them?

What to record
Put each plant back where it belongs.
Explain why it belongs there.

Explain why here:

Part 3

Sc3 Materials and their properties

Key Stage 1
National Curriculum reference: Sc3, 1c

10 What are natural materials?

What children may think

'At school we are learning about materials. At the weekend all my family went to town to choose some material for my new bedroom curtains.' Confusing, isn't it? It's even more confusing when you consider that children may use the word for a material and an object interchangeably – as we all do when we talk about a glass of water.

But children respond well to the word 'stuff'. A box may be a box, but the stuff from which it is made is cardboard. Thus you can separate an object from its material. But you can't so easily resolve which materials are natural and which are artificial or made.

What you may think

A natural material is one that you find in nature – wood, wool, stone. This is true. But we seldom use the material just as it is found. We frequently change it in some way. Does it then become 'made'? Is a gravestone a natural object? Or is it changed because it is shaped and engraved? The new bedroom curtains will be made from fabric and it helps to establish the early use of this word – as distinct from 'material'.

The accepted science

As so often in science the answer is that there is a continuum. There are, without question, materials that are used exactly as they are found. Dry stone walls are made by piling natural stones on one another, for example. The popular book *101 uses of a dead cat* suggests some ways of using a 'found' object. But most objects are changed in some way and the greater the change, the further the object is from its natural state.

Many materials are cut and shaped. Wood for example. Others may have a finish applied – paint or varnish. Some are changed with heat – clay to fired pottery, eggs to omelettes. The effects of heat usually bring about a chemical change. You can return dried 'leather' pottery to clay, but not once it has been fired. Finally there are some synthetics that have no resemblance to the original material – plastic to oil for example.

The photocopy sheets

1 Sorting real objects – e.g. toys – into natural and made.

2 The following materials are natural:

The farmer rose early.
She put on her cotton shirt and jeans.
She wore a silk scarf.
She went to the stone barn.
She put the wool fleeces from the sheep on the wood table.

Some more activities

Any classification of real materials. Opportunities arise when doing PE, modelling or cooking.

Learning outcomes

Know that some common materials occur naturally, and others are made.

What are natural materials?

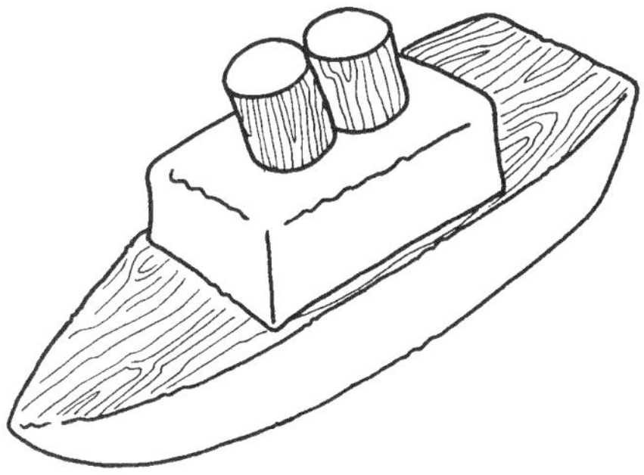

What you need
Some small toys and other objects.
This sheet and a pencil.

What to do
Sort your toys.
What materials are they made from?
Which are natural?
Which are made?

What to record
Complete this table:

Name of toy	Natural or made?	How do you know?

What are natural materials?

What you need
This sheet and a pencil.

What to do
Read this story.

The farmer rose early.
She put on her cotton shirt and jeans.
She wore a silk scarf.
She went to the stone barn.
She put the wool fleeces from the sheep on the wood table.

What to record
Underline all the natural materials in the story.

Key Stage 1
National Curriculum
reference: Sc3, 1d

11 Why use this material here?

What children may think

Why should children think twice about the uses to which we put materials? Clothes are made of cotton or wool, walls of bricks, windows of glass. This is the way the world works. And yet the selection of the right material to do a task – and in the right quantities – is at the very heart of technology.

Children faced with a challenge may make choices that are not effective in the circumstances. Paper may be chosen where card is needed; paste where a strong glue would be better; thread in place of rope. They learn to make more practical choices with experience.

What you may think

We have choices to make when it comes to meeting practical problems. What is the best potting compost for these plants? What mix of concrete will be suitable for the patio? What fabric should cover the suite? Many of those decisions may be made on financial grounds, as well as practical ones. But we recognise when unusual materials have been used. The De Lorean sports car was made from aluminium, the body unpainted. You can buy cardboard furniture, shaped to achieve the maximum strength. But you can't buy a chocolate fireguard – or a chocolate teapot – for obvious reasons. Some materials are not appropriate to the task.

The accepted science

Natural objects and materials are formed without human action. Made materials are the results of human activity. With growing understanding of the way materials behave, we can predict far more accurately what each will do in given circumstances: how long those jeans will last; whether that bridge will take the load. We can even begin to produce materials to order with the required characteristics – so-called designer materials. Thus a manufacturer can stipulate the conditions they expect the materials to experience, and the right material can be made accordingly.

The photocopy sheets

1 Children are asked to go on a materials hunt with the sheet and pencil. You will decide where they can safely go, indoors or outdoors. They record the material they see, the place it is being used, and the reason for the use, e.g. glass/window/see through. There are spaces for four answers.

2 Dozy Dora's clay car is only one of several unlikely inventions which use the wrong materials. The children are asked to explain why the materials are no good. They might invent their own improbable device.

Some more activities

Any technology activity that involves the choice of materials – to make, to fix, to mould.

Learning outcomes

Know that we chose materials for use on the basis of their properties.

Why use this material here?

What you need
This sheet and a pencil.

What to do
You are going on a materials hunt.
Your teacher will tell you where you can go.
Fill in the table.

What to record

The material	Where it is	What it does

Is this the best material?

What you need
This sheet and a pencil.

What to do
Dozy Dora made a car from clay.
The car was soft.
The wheels stuck.
Draw Dora's car.

What to record
Why are these no good?
Dreamy Dave's chocolate chair.
Dippy Dan's paper house.
Daft Daisy's rubber table.

Draw them.

Dreamy Dave's chair.

Dippy Dan's house.

Daft Daisy's table.

Key Stage 1
National Curriculum
reference: Sc3, 2b

12 How do materials change?

What children may think

The concept of materials changing is a difficult one. It's not surprising that children recognise changes when they are spectacular – a colour change, a flame, smoke – but not when they are unexciting. They may feel that stuff disappears – there was a match there, now it's gone. They may think that the product was somehow inside the original material – thus rust may be thought to ooze out of a corroding nail. They may simply think that one material has turned into another – the flour and water have turned into bread – without a word of explanation.

What you may think

There are irreversible changes that we see frequently in primary science but that are still hard to explain. Take, for example, the burning of a candle. We can all recognise the stages – solid, liquid, gas – that the wax goes through as the candle burns. But what is the stuff left at the end? Is it wax? Will it burn? Where has the lost wax gone? The fact is that the products of burning wax – water vapour and carbon dioxide – are lost – invisibly. So the change that children observe is the gradual shrinkage of the candle. It doesn't seem logical – but remember that we fill the petrol tank of the car and don't expect to clear out a load of ash next time we stop for fuel. Another puzzling change is the clarification of butter. You can melt chocolate – gently – and get the chocolate back. But if you melt butter then it clarifies and will not return to its original form. Why is this? Once again the answer is invisible. Water vapour is lost from the melted butter changing its chemical composition irreversibly.

The accepted science

Many changes are reversible. They are often changes in state – ice to water, water to ice, or dissolving, but some are chemical. However, most chemical changes are permanent. Take as an example the changes that arise when you make clay pots. The clay dries out to what potters call 'leather'. This is a physical change. The pot is shaped but if you were to put it in water, you would quickly get a lump of wet clay back again. If you fire the pot – heat it – then the change is both chemical and permanent.

The clay becomes hard pottery, and no amount of water will get the clay back again. Changes in the kitchen are more complex. Creaming flour and butter combines it – permanently. Cooking produces a completely new material.

The photocopy sheets

1 Small amounts of fats are melted in warm (not hot) water. Notice that some are permanently changed.

2 A clay pot – made by coiling the clay, perhaps – lets the water through, but the saucer does not. What has happened to the clay? Don't expect words like 'fired' but children may have ideas about an oven.

Some more activities

Any cookery activity that introduces changes in the material. This need not use heat but heat changes are permanent.

Learning outcomes

Know that some everyday materials change when heated or cooled.

How do materials change?

What you need
Some soft foods:
butter;
margarine;
chocolate;
cooking fat.
Some warm water in a bowl.
Some small sweet or cake cases.

What to do
Put a little of each food in a case.
Float them on the warm water.
Watch.

What to record
Which melted first?
Which melted most?
Which hardened first?
Which changed?

How do materials change?

What you need
Some natural clay.
Water.
A saucer.

What to do
Use the clay to make a bowl.
Put the bowl in the saucer.
Fill your bowl with water.
Watch.

What to record
What happens to the water?
What happens to the clay?
The saucer is made of clay too.
Why doesn't it soften?

Part 4

Sc4 Physical processes

Key Stage 1
National Curriculum
reference: Sc4, 1b

13 Do bulbs need two wires?

What children may think

Here's a puzzle. The television apparently has only one wire going from the wall to the set. When I switch it on, electricity seems to flow from the wall socket to the set rather like water down the pipe to the tap – and the television works. Yet when I want to light a bulb at school, I need two wires – one from the battery to the bulb, and another one back to the battery. What's going on here? What's the difference?

It's not surprising that children think in terms of a tank-and-pipe model when they first try to understand electricity. It looks as if things work because electricity flows to them through the wire, only to be used up when the television or the light is on. But you're an adult. You've wired up a plug and you know that there are two – or even three – separate wires inside the cable. So you know better – or do you?

What you may think

Well of course you need two wires. Electricity flows out of the plug or battery, does the job, and then flows back in again. The third, earth wire (if there is one) is there in case of accidents – to take the electricity – literally – to earth. Electricity flows round a circuit.

But even physics and engineering students have been known to hold incorrect ideas about the flow of electricity. A popular notion is that electricity flows from both sides of the battery, meeting at the bulb and making it light. Even those who recognise that electricity flows only one way round a circuit may think that the bulb somehow uses up some of that current, and that a weakened current crawls back to the battery. But this isn't so. Measure the current anywhere in the circuit, and the measurement will be the same. It's what the circuit does with the electricity that matters.

A reputable school textbook assures you that if you connect two bulbs in series – one after another – the first one will glow brighter than the second, because the first bulb uses up some of the electricity. Wrong. The bulbs share the electricity, and adding a second bulb to a battery circuit will dim both bulbs.

The accepted science

Forget any notion of the battery as some kind of electricity storage tank. It isn't. It is more like a pump, pushing electricity round the circuit. (And the bigger the pump, the bigger the push – potentially lethally big in the case of mains circuits.) Electricity is pushed round the circuit by the battery and the current is unchanged by lighting a bulb or spinning a motor (or working a television). It's what the electricity does as it flows through a bulb, a television or your body that matters.

In fact, electrons flow from the negative to the positive terminal of a battery (the opposite of what you might expect from the names) producing a direct current. Objects in the circuit resist that flow. Two identical bulbs in circuit resist in identical ways and will glow identically. Bulbs have a different resistance to the flow of electricity than tiny motors – which is why a motor may whirr while a bulb in the same circuit fails to light – the bulb is like a wide corridor to the electrons, and the motor a narrow one. But only a switch will stop the flow by breaking the circuit.

The photocopy sheets

1 There has to be a full circuit for the bulb to light. Most children do not know that the side and end of the bulb are the connection points. Pressing the end of the bulb to the end of the battery, and linking the other end of the battery to the side of the bulb with a single wire completes the circuit.

2 The bulb will light where there is a full circuit. One wire must touch the side of the bulb, one wire the end.

Some more activities

Nothing will beat practical experience with bulbs, wires and batteries.

Learning outcomes

Know that a full circuit is needed for electricity to flow.

Do bulbs need two wires?

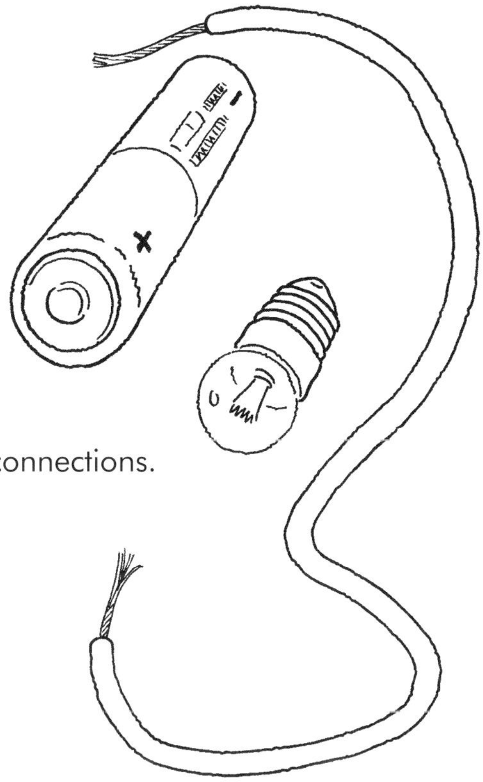

What you need
A bulb.
A battery.
One wire.

What to do
Light the bulb using only one wire.

What to record
It can be done.
You must find a way of making two connections.

Draw your answer.

Do bulbs need two wires?

What you need
This sheet and a pencil.

What to do
Look at these pictures.
The bulb will sometimes light up.

What to record
Tick the ones where the bulb will work.
Cross the ones where it will <u>not</u> work.

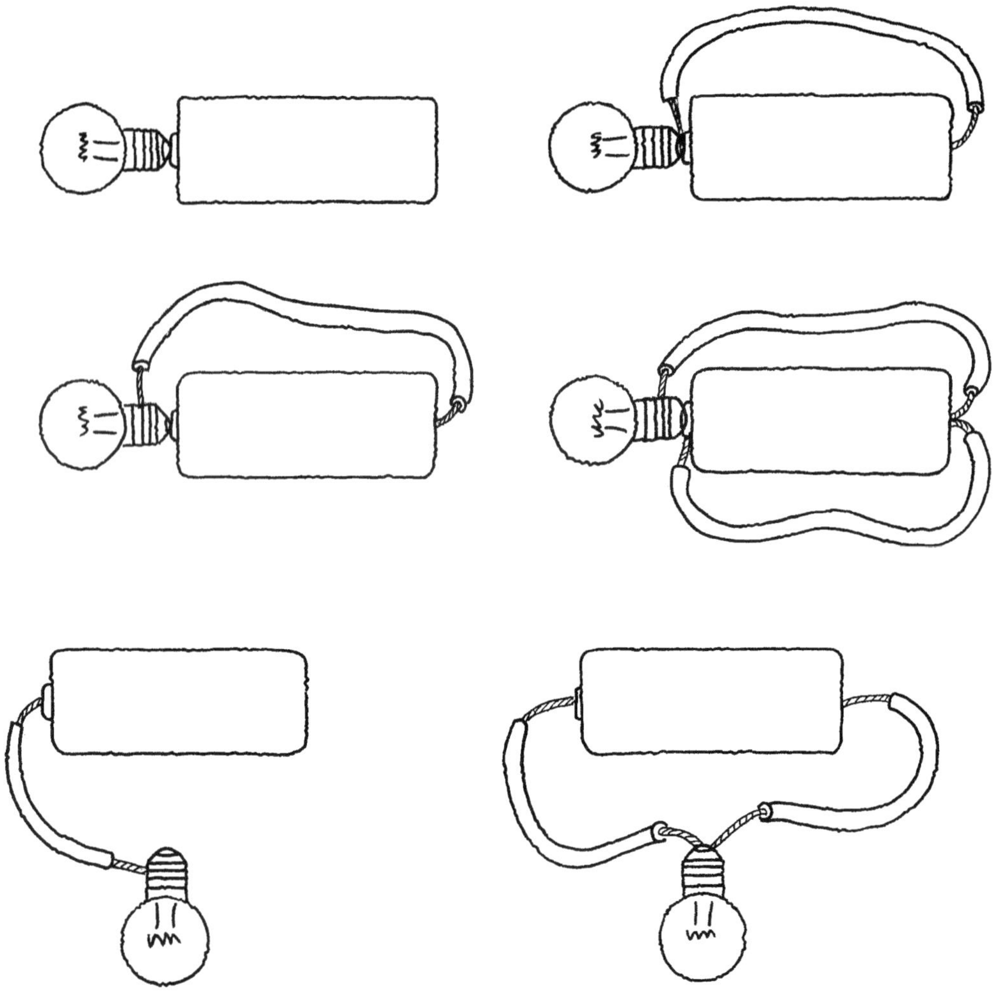

Key Stage 1
National Curriculum reference: Sc4, 2a, 2b, 2c

14 How do you recognise a force?

What children may think

It's not very nice to force someone to do something, is it?
My dad forces rhubarb with an old bucket.
I'm really strong. I've got lots of force.
My sister's in the police force.

We use the word 'force' in so many different contexts and in so many different ways that it is not surprising that children have difficulty in coming to terms with the idea. Even when they have grasped what is meant by a force in science, they may still feel that forces are somehow active and deliberate, as if they have a mind of their own.

What you may think

'Forces' has to be the all-time pet hate of teachers of primary science. Recent changes in the curriculum simplifying the requirements can only be welcomed. Even when you have grasped that a force is a push or a pull you are left with situations where there are forces acting but nothing is moving. What is going on?

The accepted science

A force is a push or a pull. There are fancy types of pushes and pulls – like twists for example. But basically forces are pushes and pulls and we can recognise plenty around us – cars pulling caravans, people pushing prams. But there are still forces operating when things are standing still. We can come to terms with this when something is almost moving as when two Sumo wrestlers are pushing shoulder to shoulder, or two tug-of-war teams are pulling equally. We can see how a slight change will tip the situation to movement.

But a book on a table? Are there forces involved? Well yes. If you took the table away, the book would fall. It is subject to the pulling force of gravity. But how can a table push? Take a metre rule and put one end on a table edge. Hold the other. Balance a book in the middle. The ruler bows. It is easy to imagine that the bending ruler is pushing back on the book. Shorten the ruler. Now it doesn't bow as much. But you can still see it pushing. Shorten it still more. You can no longer see it bowing. But it is still pushing back. Even stationary things exert a force.

The photocopy sheets

1 Children are to spot and circle the forces.

2 Children stick Post-its saying 'push' or 'pull' to classroom objects. They must not try objects that are dangerous or electrical devices.

Some more activities

In PE try some safe push and pull activities – especially those like tug of war or leaning on each other that balance the forces.

Learning outcomes

Know that a force is a push or a pull, and that pushes and pulls speed things up, slow things down, and change their shape or direction.

How do you recognise a force?

What you need
Your pencil and this sheet.

What to do
Look for pushes and pulls.

What to record
Circle the forces in action.

How do you recognise a force?

What you need
Some Post-its.
Your pencil.

What to do
Look around the room.
Spot things that you might push or pull.

What to record
Stick a 'pull' label on things you pull.
Stick a 'push' label on things you push.

Key Stage 1
National Curriculum
reference: Sc4, 3a

15 What is a light source?

What children may think

The word doesn't help. It's both a verb and a noun. 'Light the light, give us some light' is a clumsy sentence, but not incorrect. As a noun, it can have two meanings – light as a source (a light bulb) and light as a phenomenon (light filled the room). Never mind the other uses of the word – a light load, light-headed, give us a light, light comedy, light green and so on.

What you may think

You – like most children – are unlikely to have experienced total darkness. It confirms – as no other experience does – that we need light to see. Even a child tucked up in the dark can still see. There is light through the curtains, or under the door from the landing.

But what are light sources? Children may think that they include the Sun, lamps, the Moon, candles, mirrors and torches. Only four of these are sources. The Moon and mirrors both reflect light.

The accepted science

We see objects because light from a source is reflected from the object. If you search for something in the cellar with a torch, your eyes follow the torch beam as you look around – you don't point the torch at the floor while looking at the ceiling. this seems obvious now, but a lot of scientists have thought otherwise, throughout history. It was the Arab scientist Ibn al Alhazen who finally concluded what we now understand – that light travels and that our eyes are receptors of light – not transmitters of magic beams.

The photocopy sheets

1 Light sources and light reflectors. Children choose and name each.

2 How do we see? Children are offered three choices. Light from a source, reflected from an object and into the eye is the correct choice.

Some more activities

Try using a torch in a dark room or dark corner. We can only see things when light from a source is reflected from them.

Learning outcomes

Know that light comes from a source and that we need a light source to see.

What is a light source?

What you need
This sheet and a pencil.

What to do
Look at the objects on this sheet.
Some are sources of light.
Some reflect light.

What to record
Write 'source' next to each object that produces light.
Write 'reflector' next to each object that reflects light.

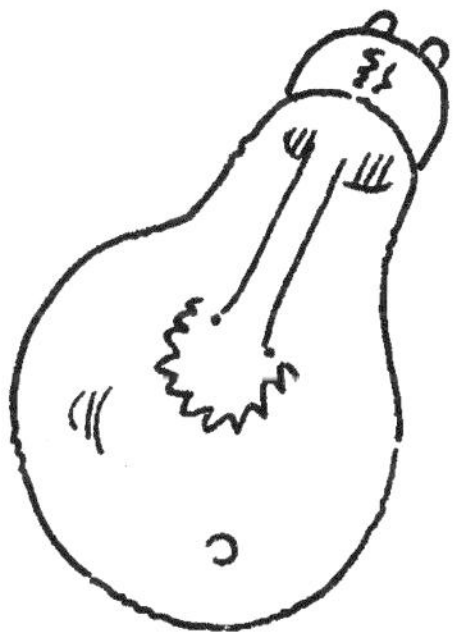

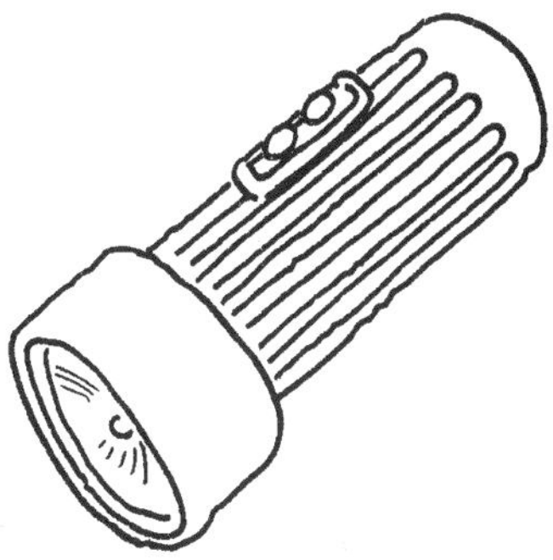

How do we see?

What you need
This sheet and a pencil.

What to do
Look at these pictures.
One shows how we see.

What to record
Draw a ring round the correct picture.
Explain what is happening.

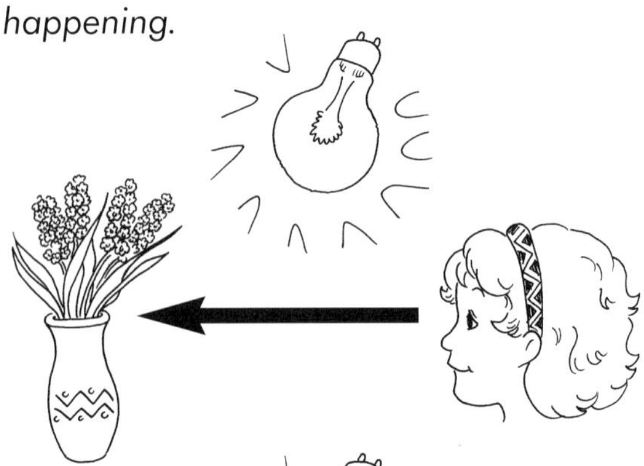

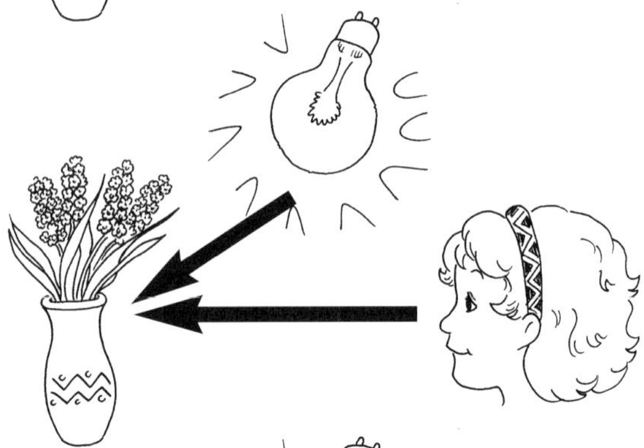

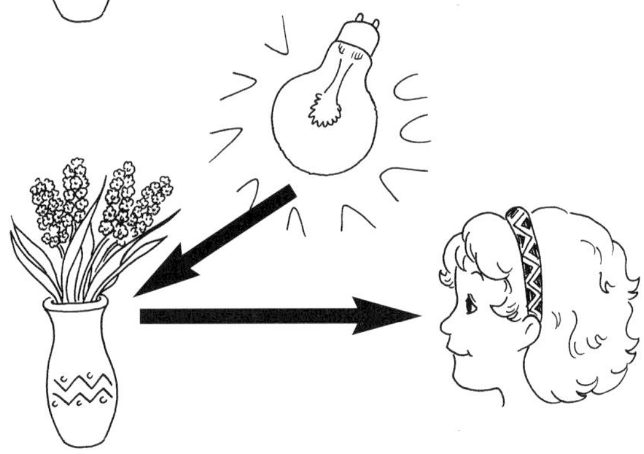

Key Stage 1
National Curriculum reference: Sc4, 3d

16 Why are you getting quieter?

What children may think

I hear the playtime bell because I'm standing in front of it. It's a good thing that there is nothing in the way, or I might not hear it. I've noticed that the bell sounds louder when I'm standing close to the teacher. But when I sing louder in hymn practice, my teacher says 'louder, not higher'. It's all very confusing.

What you may think

You probably have a clear idea of sound transmission. That when you ring the playground bell, children all over the playground can hear it, not just those in front of you. That the sound fades with distance – in all directions (and up and down). That louder is different from higher, and softer from lower. That sound needs a medium to travel – usually air. "In space, no-one can hear you scream". But you may be baffled by the unlikely assertion that sound travels faster and better though liquids than through gases, and through solids than through liquids. After all, closing the window shuts out the noise from the playground.

The accepted science

Sound travels through a medium as a series of pressure waves. These waves are generated by something vibrating. Often these vibrations are invisible. Most commonly sound waves travel through the air to you. Your ear is a receptor, responding to the changes in pressure and sending messages to the brain, which interprets the vibrations as Beethoven's Fifth Symphony or a cry for help. Conveniently, there is air in contact with both the sound generator – violin or voice – and your ear. This is called a good coupling. The sound coupling is not so good when you put your head underwater, or the sound is being generated the other side of a wall – though you can improve matters with the wall by pressing a glass against it and your ear to the glass.

Since sound travels by pressure on molecules, molecules that are close together – in a liquid – or really close together – in a solid – transmit it better than molecules that are far apart – in a gas. No molecules – no sound. No sound in space where there are few molecules. So you can hear a train coming from a long way away when the solid rails begin to sing long before you can hear the train through the air.

The photocopy sheets

1 Children are asked to record what happens to sound as you get further from it. They will find that sound travels in all directions, diminishing as it does so.

2 The ear trumpet focuses the sound on the ear and improves hearing.

Some more activities

Ask the children to spread out from the sound. Use flags to mark the points at which children stop to listen. Do they hear it getting quieter?

If you have safe stairs up or down, try the same exercise up and down rather than further away. Do children notice that the sound gets quieter with distance?

Learning outcomes

Know that sound travels, getting fainter, and that we hear it with our ears.

Why are you getting quieter?

What you need

Something to make a sound - a bell or a drum.
A safe open space - the playground.

What to do

Ask a friend to make some sound.
Walk away from them.
Listen.
Where does the sound go?

What to record

Draw a picture of your friend making a sound.
Show what happens to the sound.

How can I make you louder?

What you need
An ear trumpet, made from a cone of stiff paper.
Something to make sounds.
A safe open space.

SAFETY: Do not put the point of the ear trumpet into your ear.

What to do
Ask a friend to go into the open space and make a sound.
See if you can hear them more clearly with the ear trumpet.

What to record
Draw and write to show how the ear trumpet works.